Editor
Sarah Beatty

Editorial Project Manager
Mara Ellen Guckian

Editor-in-Chief
Sharon Coan, M.S. Ed.

Illustrator
Sue Fullam

Cover Artist
Brenda DiAntonis

Art Manager
Kevin Barnes

Art Director
CJae Froshay

Imaging
James E. Grace
Ralph Olmedo, Jr.

Product Manager
Phil Garcia

Publisher
Mary D. Smith, M.S. Ed.

Measurement

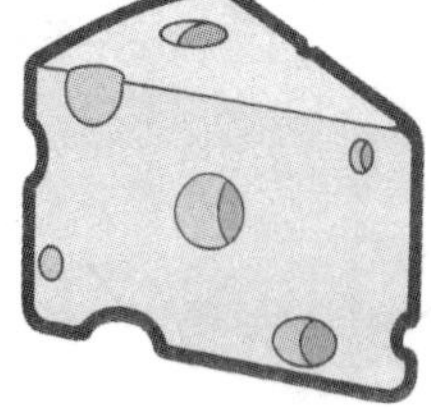

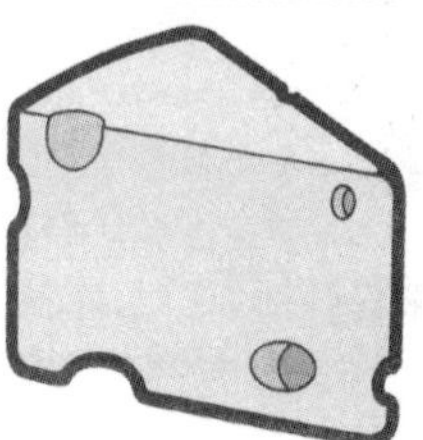

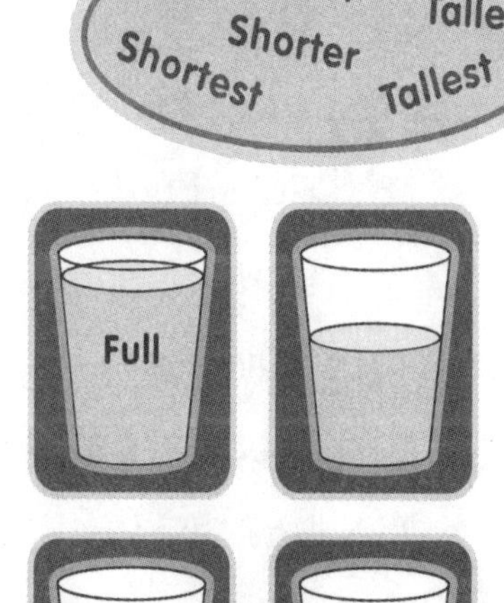

Author

Amy DeCastro, M.A.

Teacher Created Resources, Inc.
6421 Industry Way
Westminster, CA 92683
www.teachercreated.com.
ISBN-0-7439-3232-3

Reprinted, 2005
Made in U.S.A.

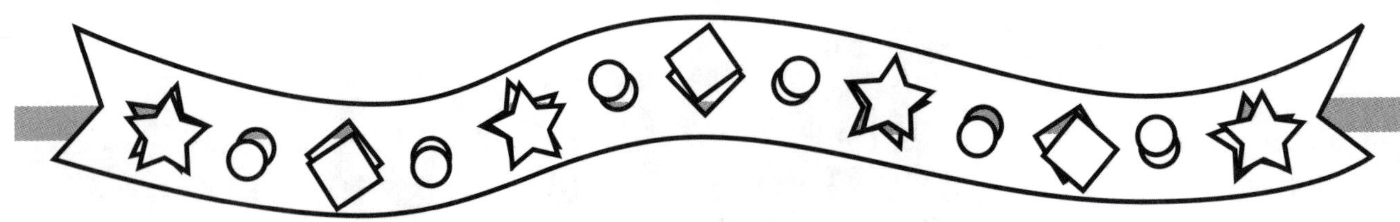

Table of Contents

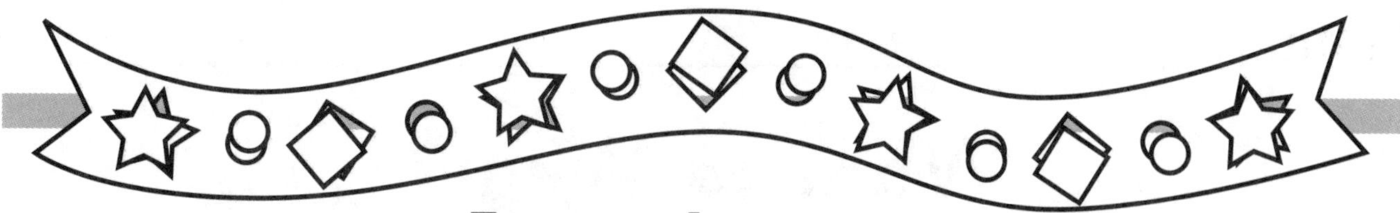

Introduction

Getting children ready for academic success starts in the early years. This poses special challenges because it is when children's attitudes towards school and learning are shaped. This workbook series' ultimate purpose is not only to promote children's development and learning, but to make that journey an enjoyable experience. Young children need lots of repetition and directions that are simply worded. The activities need to be enjoyable and visually stimulating. The series was developed with exactly that in mind. Each activity book was designed to introduce young learners to new concepts and to reinforce ones already learned. Through practice, children learn the many skills they will need for school and later life.

Measurement is one of the most widely used applications of mathematics. Early measurement experiences help students make connections between spatial concepts and numbers, and gives them opportunities to compare objects, and to count. Measurement activities can simultaneously teach important everyday skills, strengthen students' knowledge of other important topics in mathematics, and develop measurement concepts and processes that will be formalized and expanded in later years.

You will be delighted as you watch your children discover how interesting and fun learning can be all year long with the gradual sequence of one-page, easy-to-follow, enjoyable practice activities. They are great for enrichment, classroom practice, tutoring, home schooling, or just for fun.

With *Measurement,* young learners will gain basic beginning knowledge of estimating, classifying, describing, and arranging objects using comparative language in conjunction with length, size, area, weight, volume, time, and temperature. Within the activity book, students will be exposed to five areas of measurement:

Temperature—Students will identify hot and cold and thermometer reading.

Time—Students will be exposed to math vocabulary and comparisons such as before/after, and AM/PM. They will be exposed to sequencing of events, comparing duration of time, and identifying clocks as instruments to tell time.

Linear—Students will measure an object's length using non-standard units, compare one item to several units, and compare several units to one item. They will sequence objects by size and identify a ruler as an instrument for measuring.

Volume—Students will measure volume using non-standard units, identify volume vocabulary such as *empty* and *full* as well as *more* and *less.*

Mass and Weight—Students will measure mass and weight using non-standard units, estimate and compare weight of different objects, identify *heavy* and *light*, and put objects in order by weight.

Name ______________________________

What to Wear?

Directions: Look at the items of clothing. Cut out the articles of clothing and decide in what type of weather they would be worn. Glue them in the appropriate boxes.

Name______________________________

Hot Temps

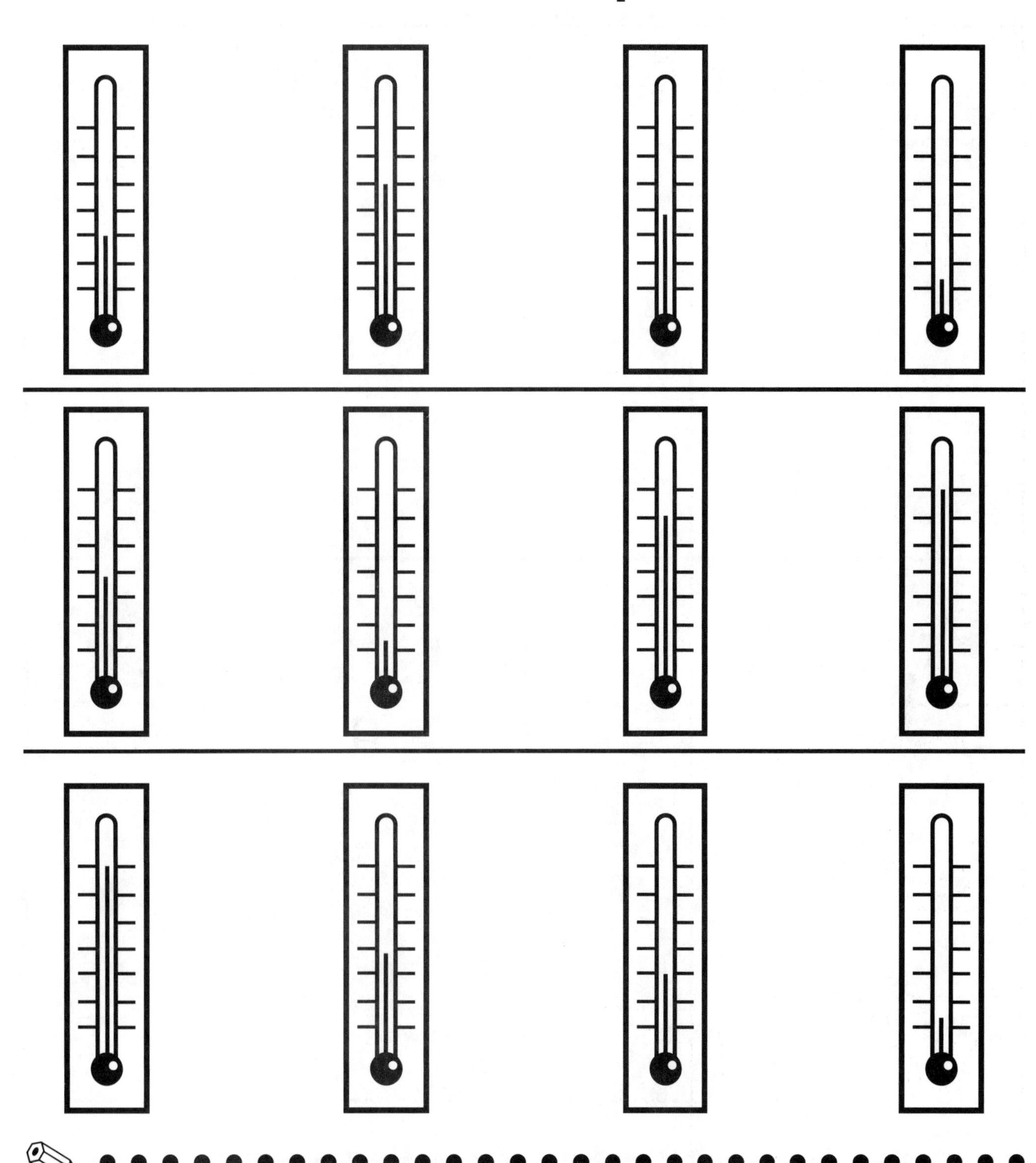

Directions: The taller the line on a thermometer, the hotter the temperature. Circle the thermometer in each row that shows the hottest temperature. Cross out the thermometer with the coldest temperature in each row.

Name ______________________________

Brrr, Cold!

Directions: The weather is cold outside. Draw two things above that you need to wear when it's cold.

Name ______________________________

Hot! Don't Touch!

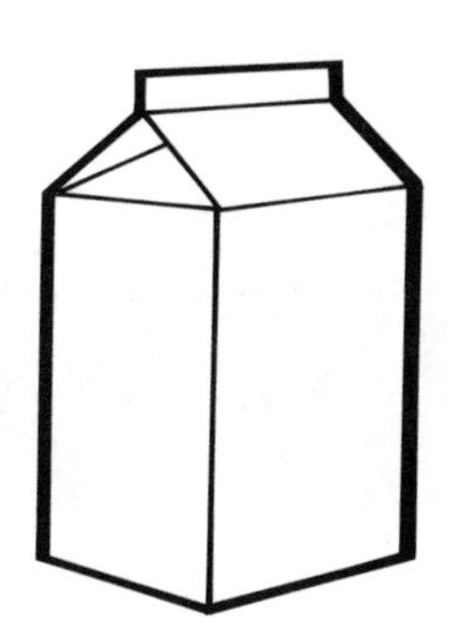
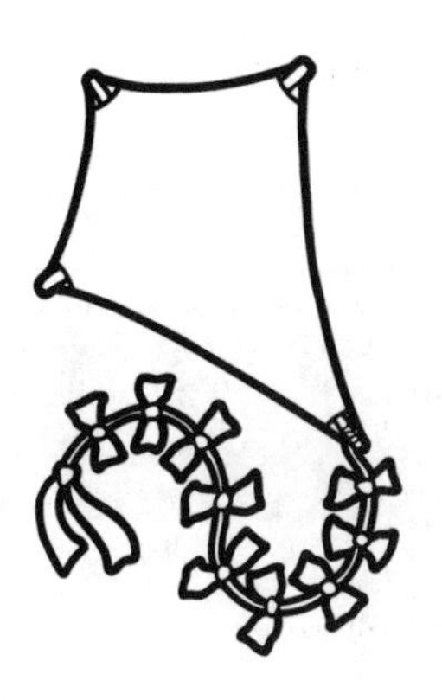

Directions: Look at the pictures above. Cross out the item in each row that is too hot to touch. Color the items that are safe to touch.

Name ________________________________

Weatherman

Sunday	Monday	Tuesday	Wednesday

Thursday	Friday	Saturday

Directions: Keep track of the weather at the same time each day for a week.

- Draw a ☼ if it's sunny that day.
- Draw a ☁ if it's cloudy that day.
- Draw a 💧 if it's rainy that day.
- Draw a ✳ if it's snowy that day.

Name____

Day and Night

Night

Day

Directions: In the first box, draw a picture of something you do at night. In the second box, draw a picture of something you do during the day.

Name ______________________________

How Long Will It Take?

Directions: In each box, color the task that would take the most time to do.

Name ______________________________

Time for School

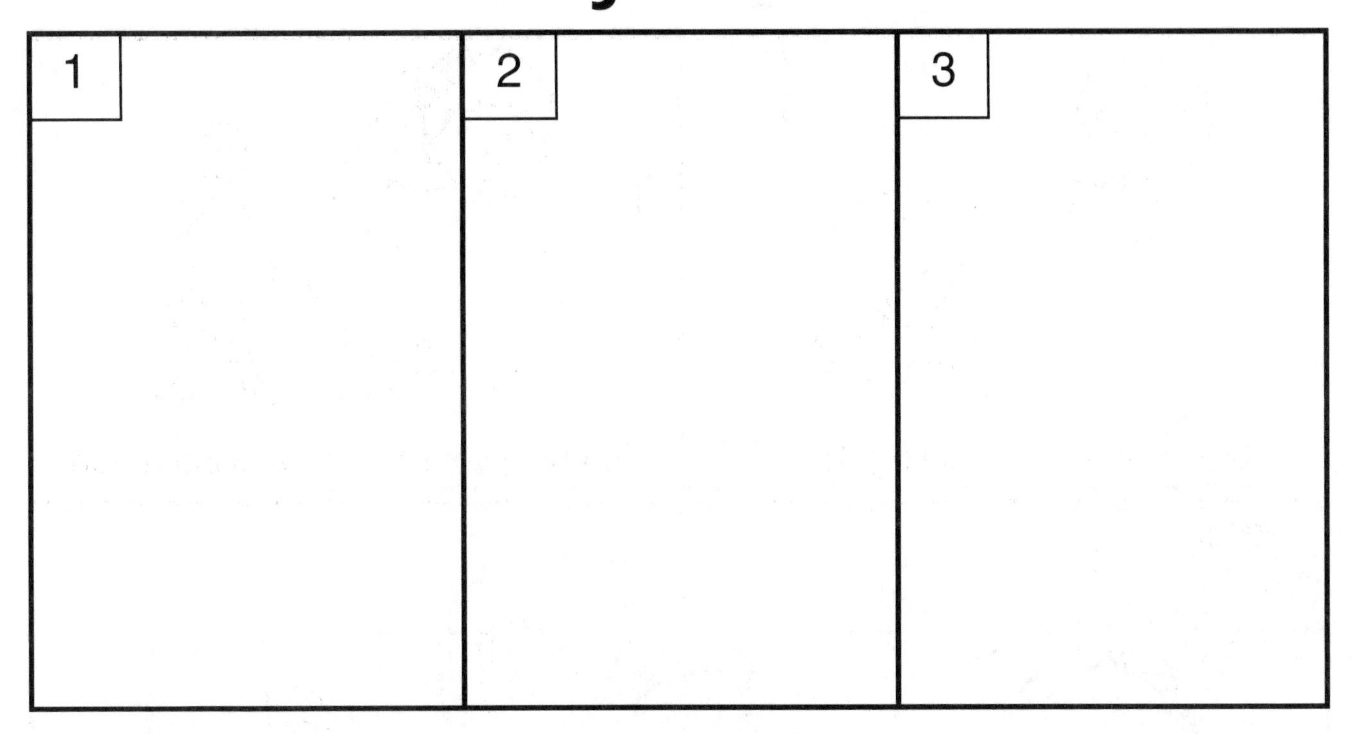

Directions: Cut out the pictures of the boy going to school. Glue them to the numbered boxes to put the events in order.

Name ______________________________

First, Next, Last

Directions: Number the events 1, 2, or 3 showing the order in each row of pictures.

Name ________________________________

What Comes Next?

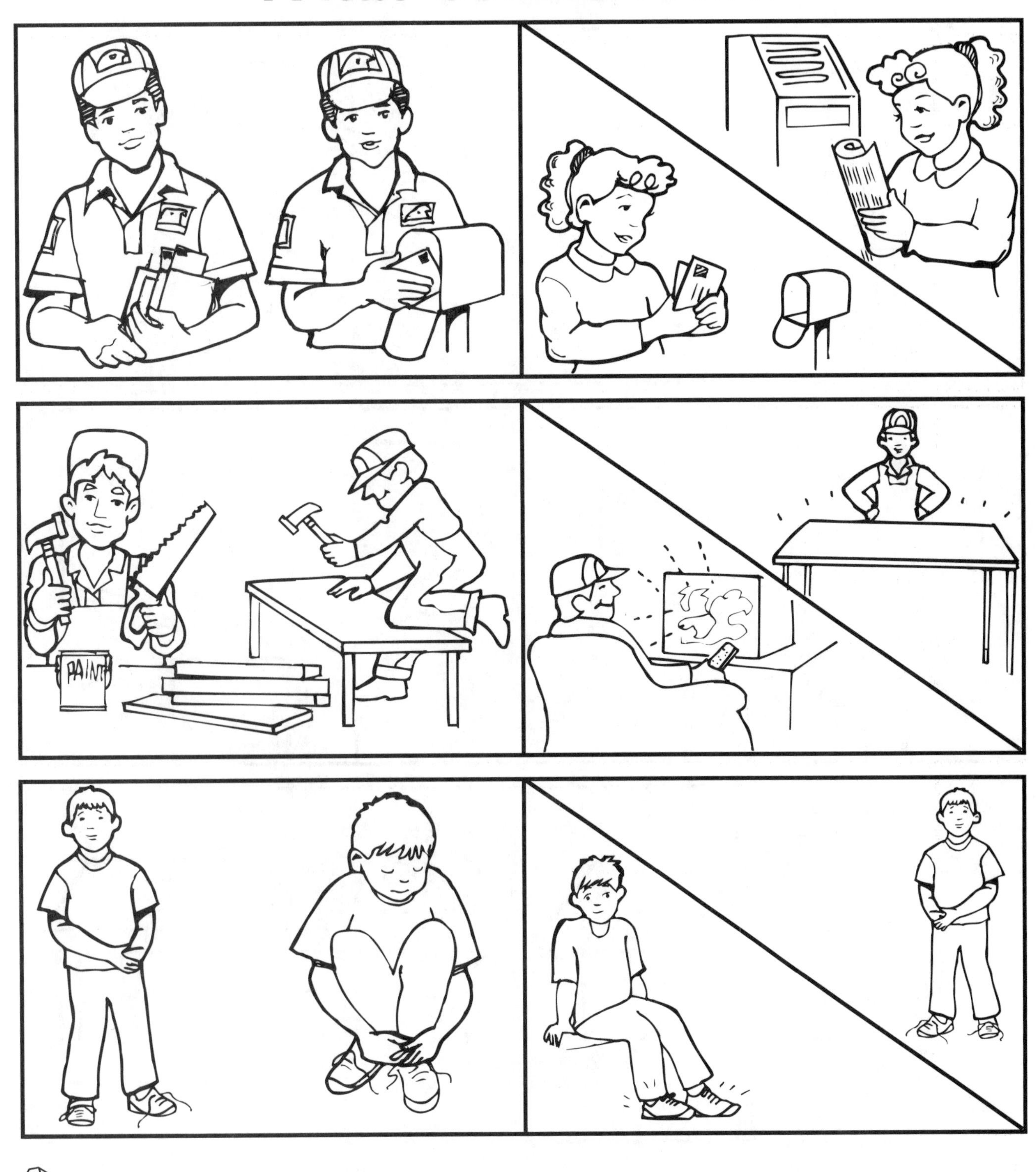

Directions: Look at each row of pictures above. Choose the picture from the box with the diagonal that comes next and color it. Cross out the other picture.

Name ______________________________

Be My Guest

Directions: Look at each set of pictures. Put the letter "B" in the box if it appears to be before the guest has arrived. Put an "A" in the box if it appears to be after the guest has been there.

Name____

Tick Tock

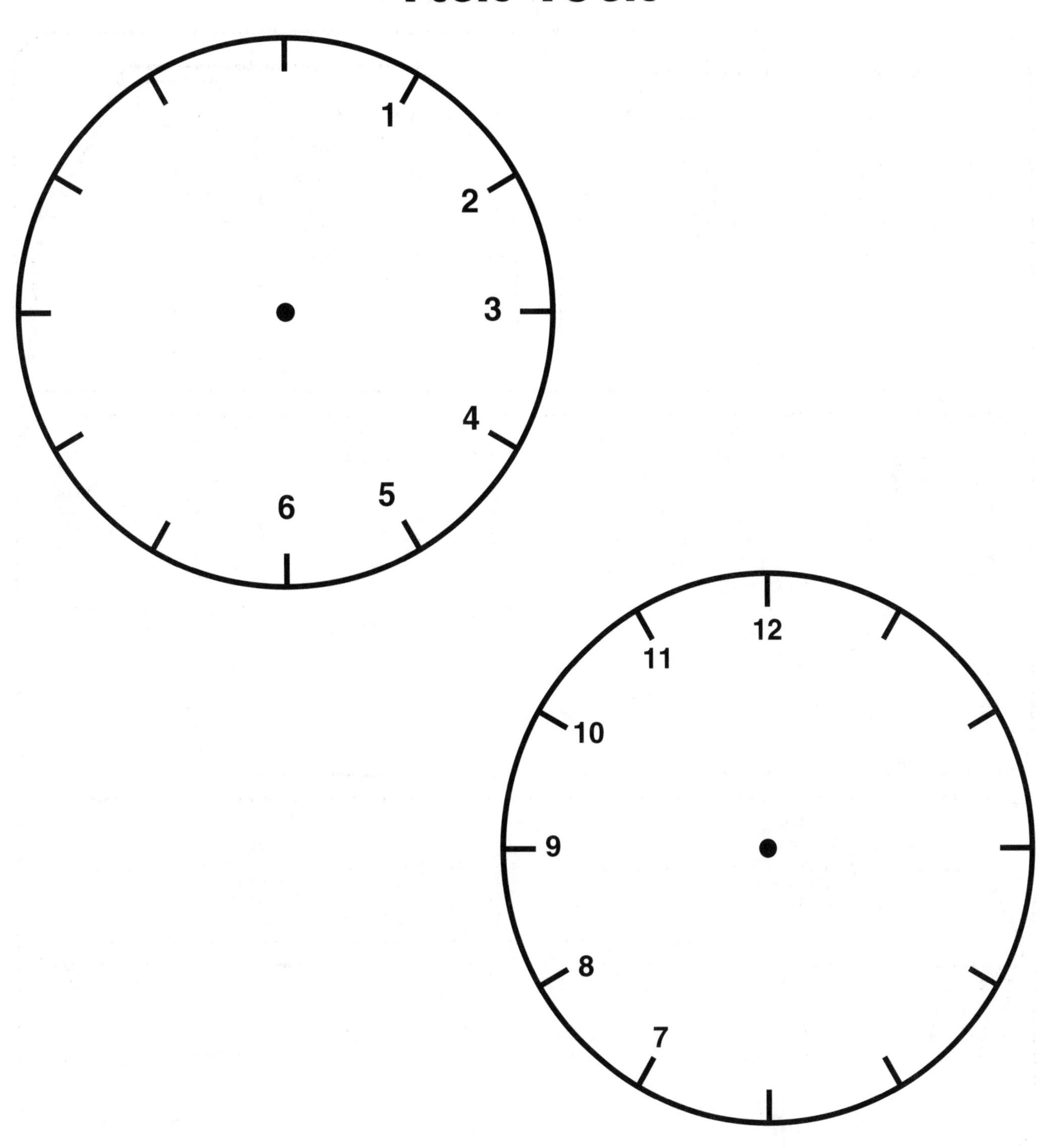

Directions: Look at the clocks above. Fill in the missing numbers. What else does a clock need? Can you put an hour hand and a minute hand on each clock? The hour hand is always shorter than the minute hand.

Name____

What Time Is It?

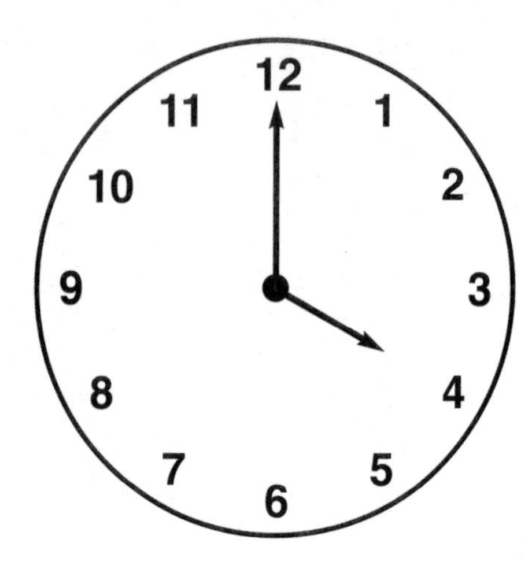

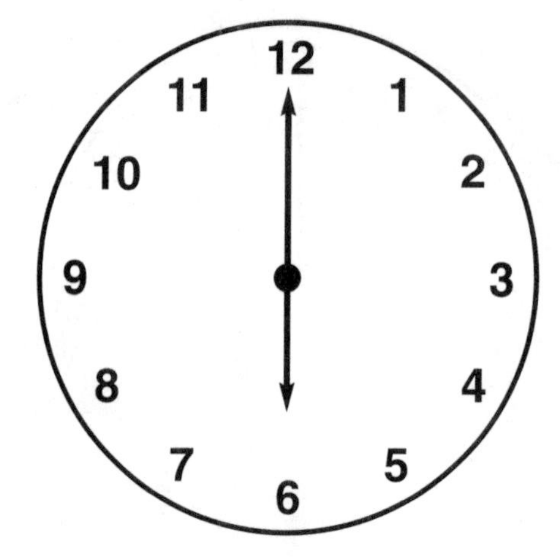

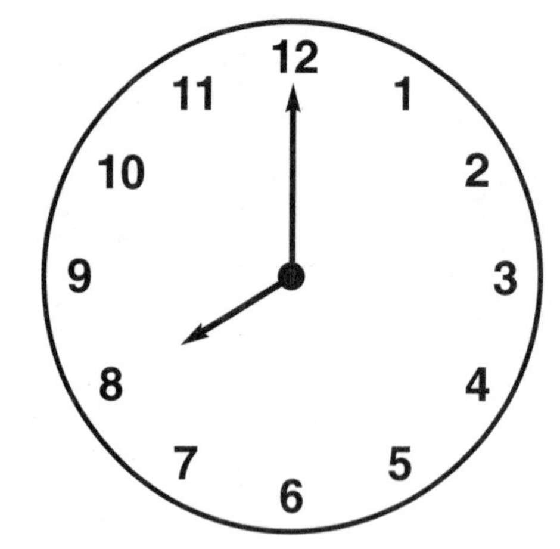

Directions: Look at each clock. The minute hand (longer hand) is on the 12 on each clock. Look at the hour hand (shorter hand) on each clock. The hour hand will tell you what time it is. Circle the number that the hour hand is pointing to on each clock. Can you tell what time it is?

Name ______________________________

Telling Time

✎ •

Directions: Look at the number the small hand (hour hand) is pointing to on each clock. Circle the number. Write that number on the line below each clock to tell the time.

Name________________________________

Don't Be Late

Directions: There's so much to do today. Make the hour hand (short hand) on each clock to show the time for each activity. Color the pictures.

Name ______________________________________

Match the Time

Directions: Draw a line to the matching time.

Name ________________________________

Jack and the Beanstalk

Directions: Draw green leaves on the *tallest* beanstalk.

Name ____________________

Firefighters

- Color the firefighter's hair that is the *longest* yellow.
- Color the firefighter's hair that is the *shortest* brown.
- Color the firefighter's legs that are the *longest* blue.
- Color the firefighter's legs that are the *shortest* red.
- Color the firefighter's hose that is the *longest* orange.
- Color the firefighter's hose that is the *shortest* purple.

Directions: Look at the two firefighters above. Follow the coloring directions.

Name ________________________________

Stack Up

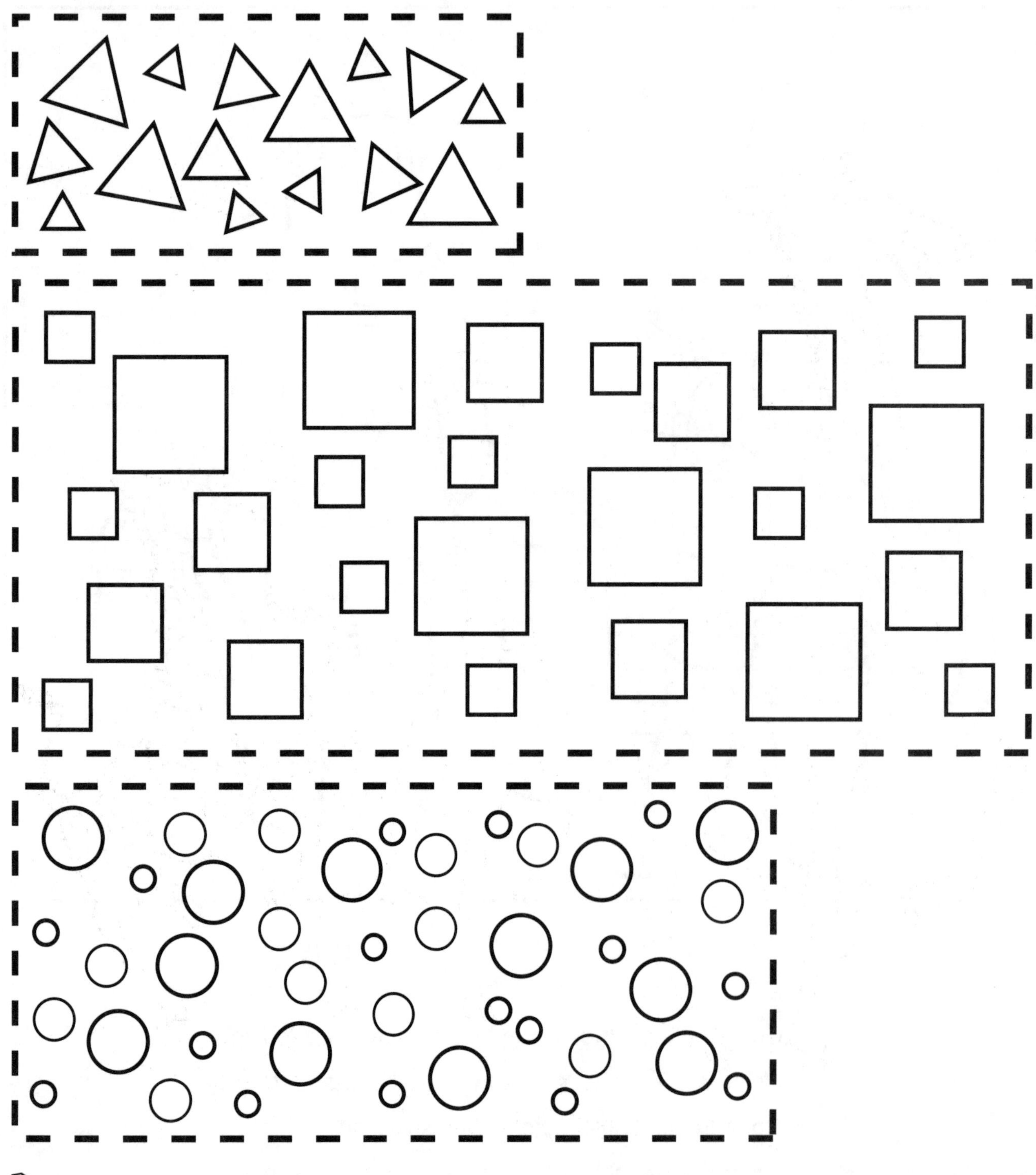

Directions: Cut out each block on the dashed lines. Stack them in order from the largest at the bottom to the smallest at the top. Glue them to a separate sheet of paper. Decorate each block.

Name ____________________

Giraffes

Directions: Look at the three giraffes above. Draw a leaf in the mouth of the giraffe that has the longest neck. Color the spots on the giraffe that has the shortest neck.

Name ______________________________

Take the Shortcut

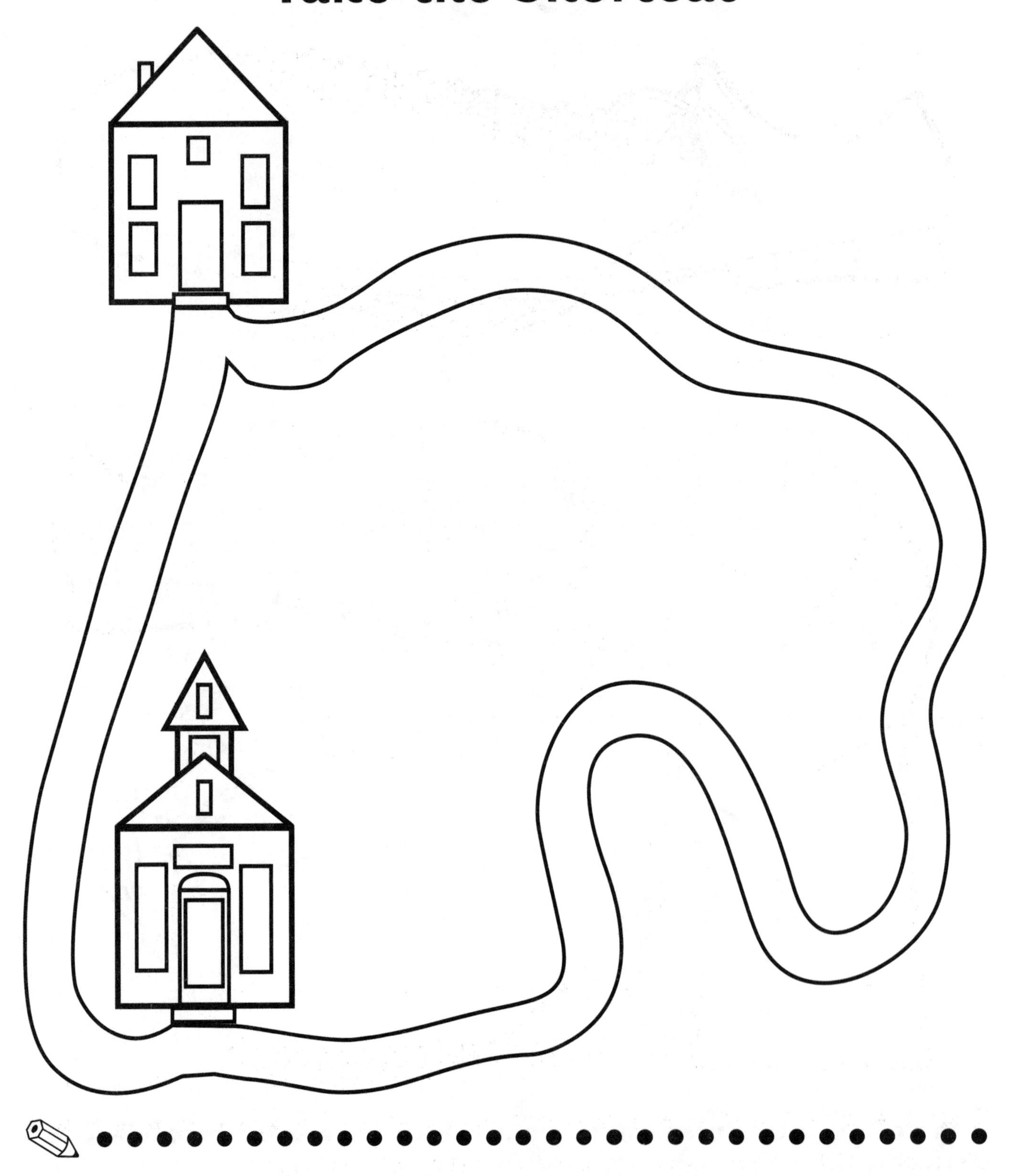

Directions: Look at the different paths that can be taken to get to school. Use your red crayon to trace the path that is the longest. Use your green crayon to trace the path that is the shortest. Color the picture. Add details.

Name ______________________________

My Shoe

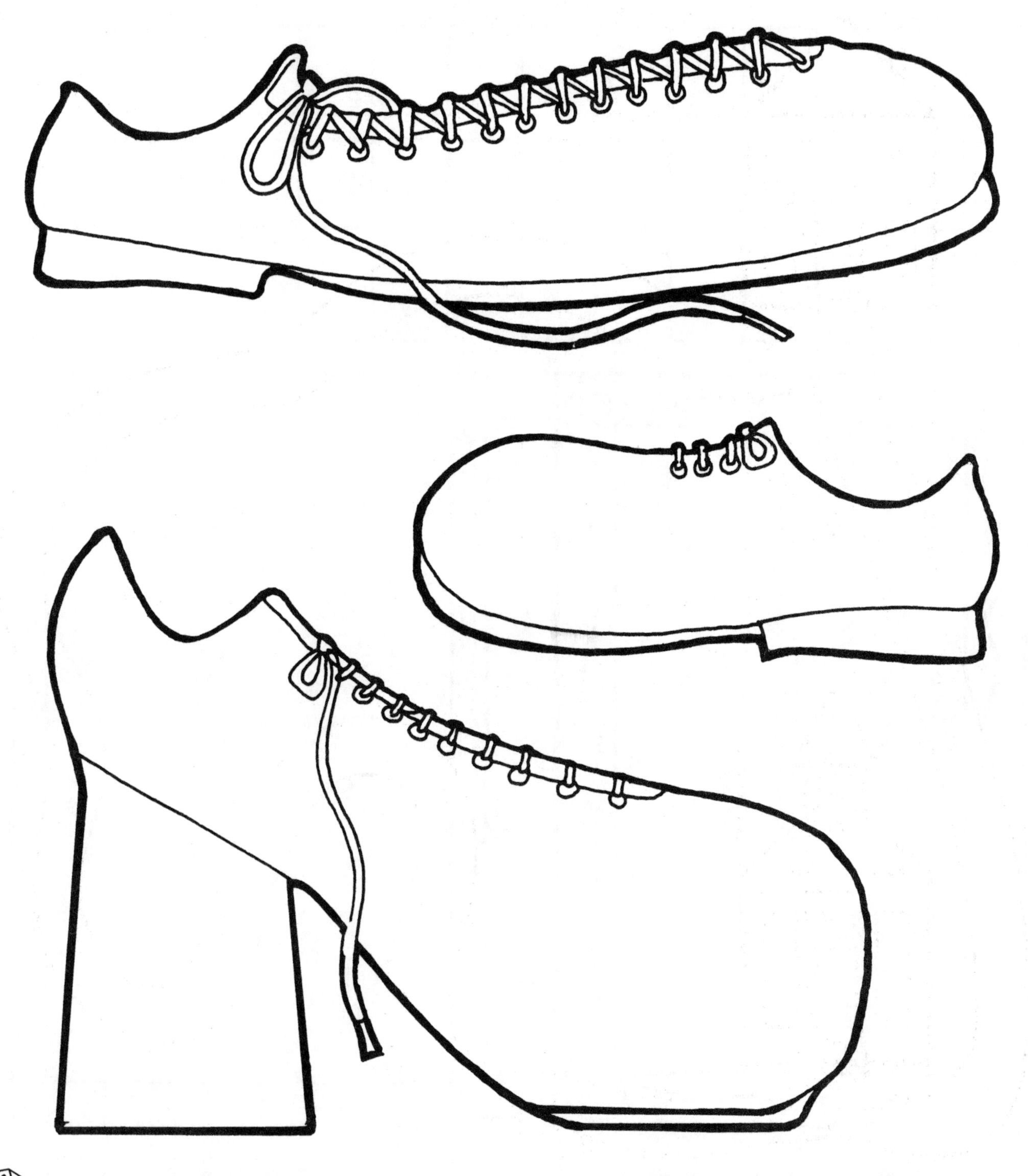

Directions: Listen to my clues to figure out which shoe belongs to me.

- My shoe is not the longest. Cross out the longest shoe.
- My shoe does not have the shortest laces. Cross out the shoe with the shortest laces.
- My shoe has the highest heel. Circle my shoe.

Name________________________________

Name It

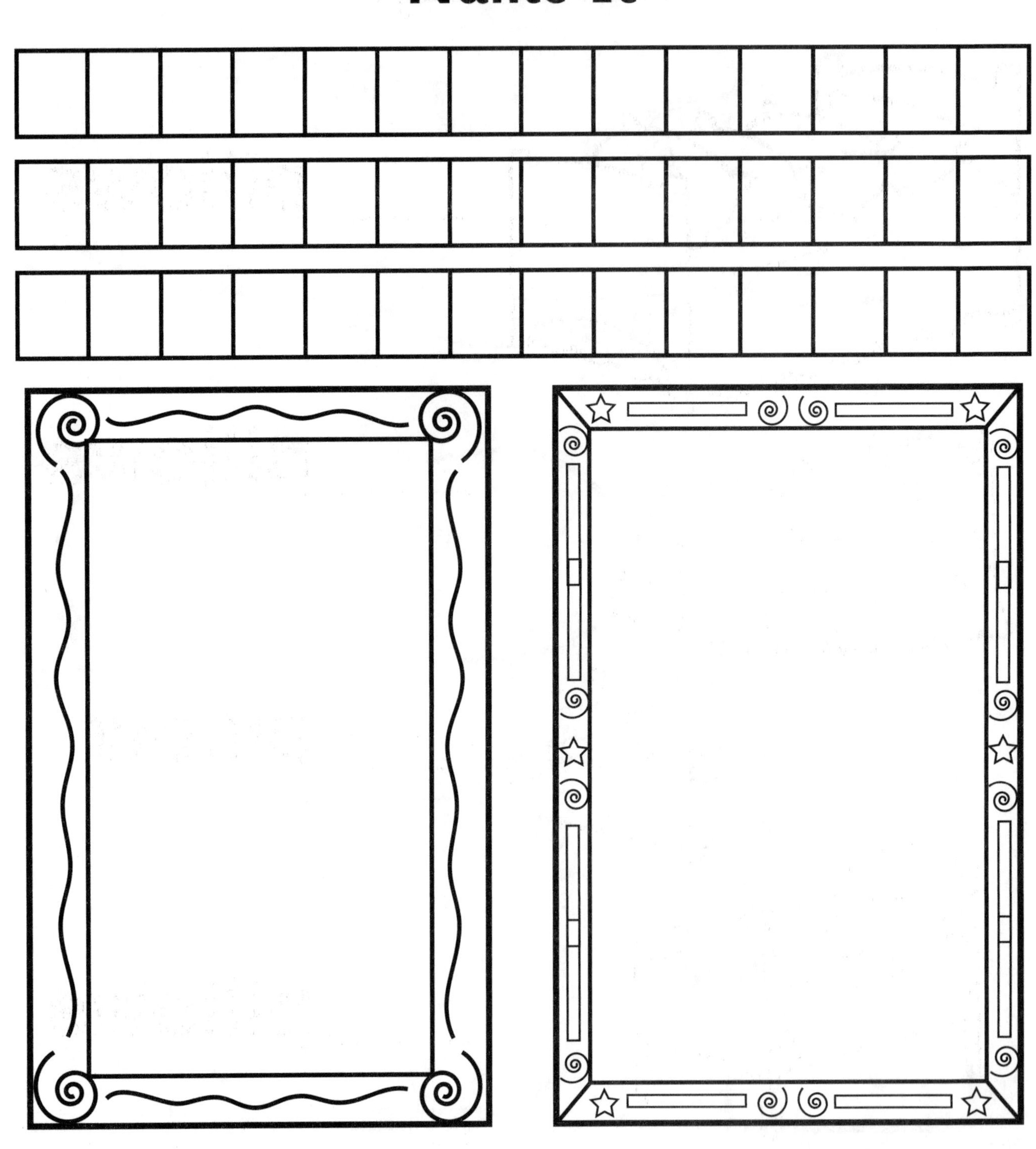

Directions: Write each letter of your name in the first row of boxes. Have a friend write each letter in his or her name in the next row of boxes. Have another friend write each letter in his or her name in the last row of boxes. Draw a picture of the person who has the longest name in the first frame. Draw a picture of the person with the shortest name in the second frame.

Name ____

How Long?

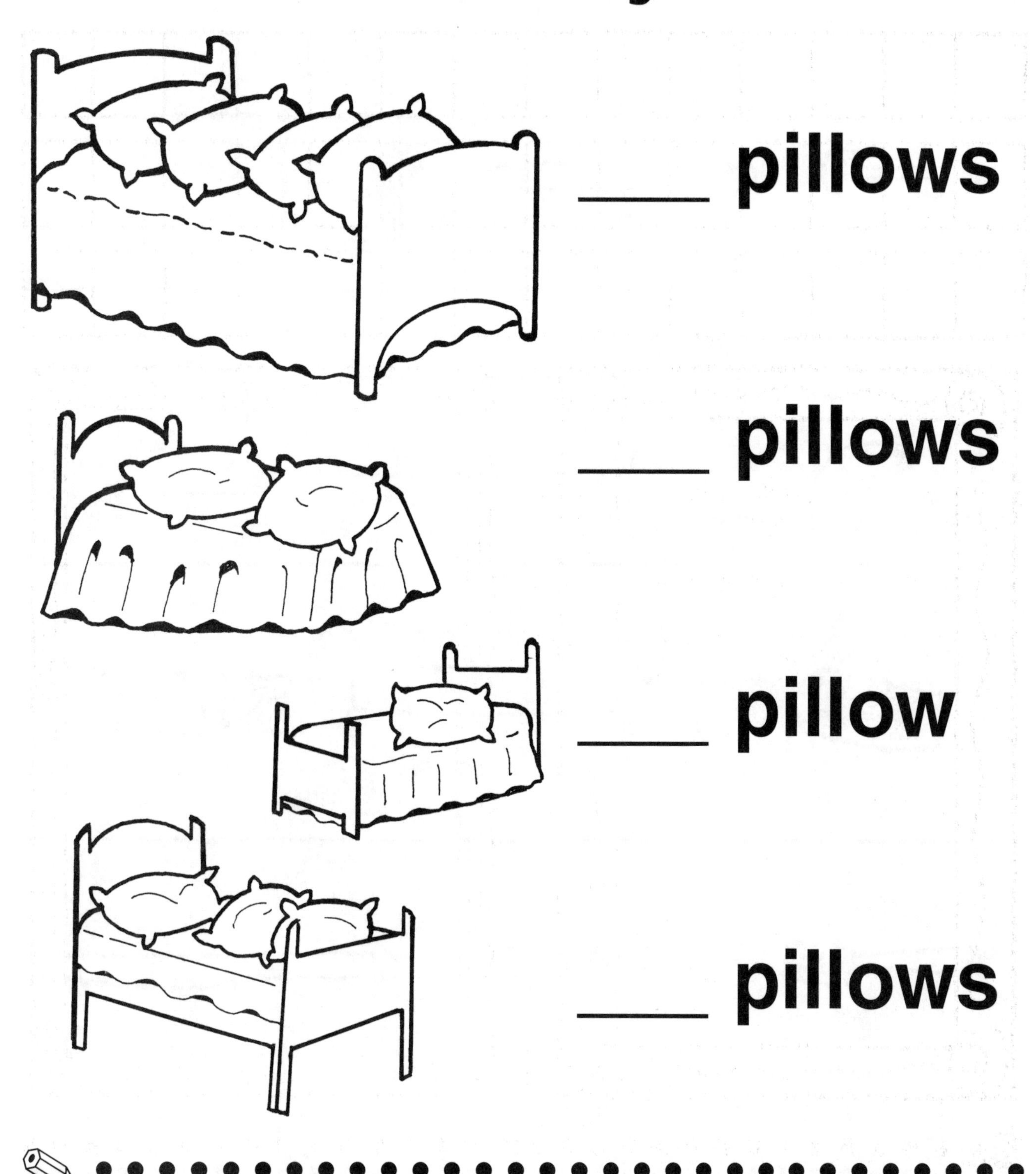

Directions: Count the pillows on each bed. Write the number on the line next to each bed. Circle the bed with the most pillows. That is the longest bed.

Name ______________________________

Same or Different?

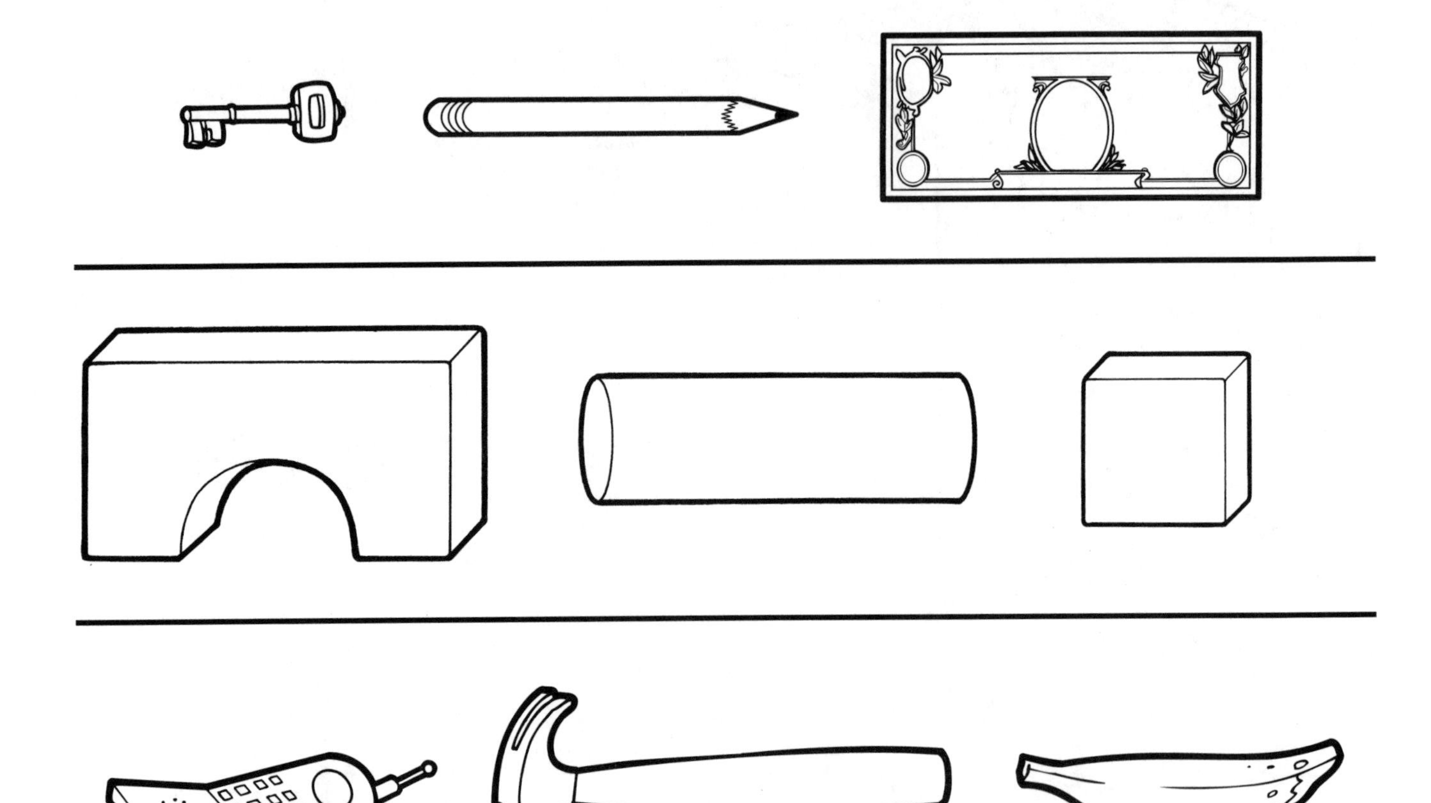

Directions: Although the three items in each row are different, two of them are the same length. Color the objects that are the same length in each row. Cross out the item that has a different length in each row.

Name ____________________________

Pet Shop

Directions: Cut out the three groups of animals. Glue each animal in order from smallest to largest on a separate sheet of paper.

Name ______________________________

Choo, Choo

_______ **long**

_______ **long**

_______ **long**

Directions: Each section of train track is numbered. How many sections long is each train?

Name________________________________

What Is Missing?

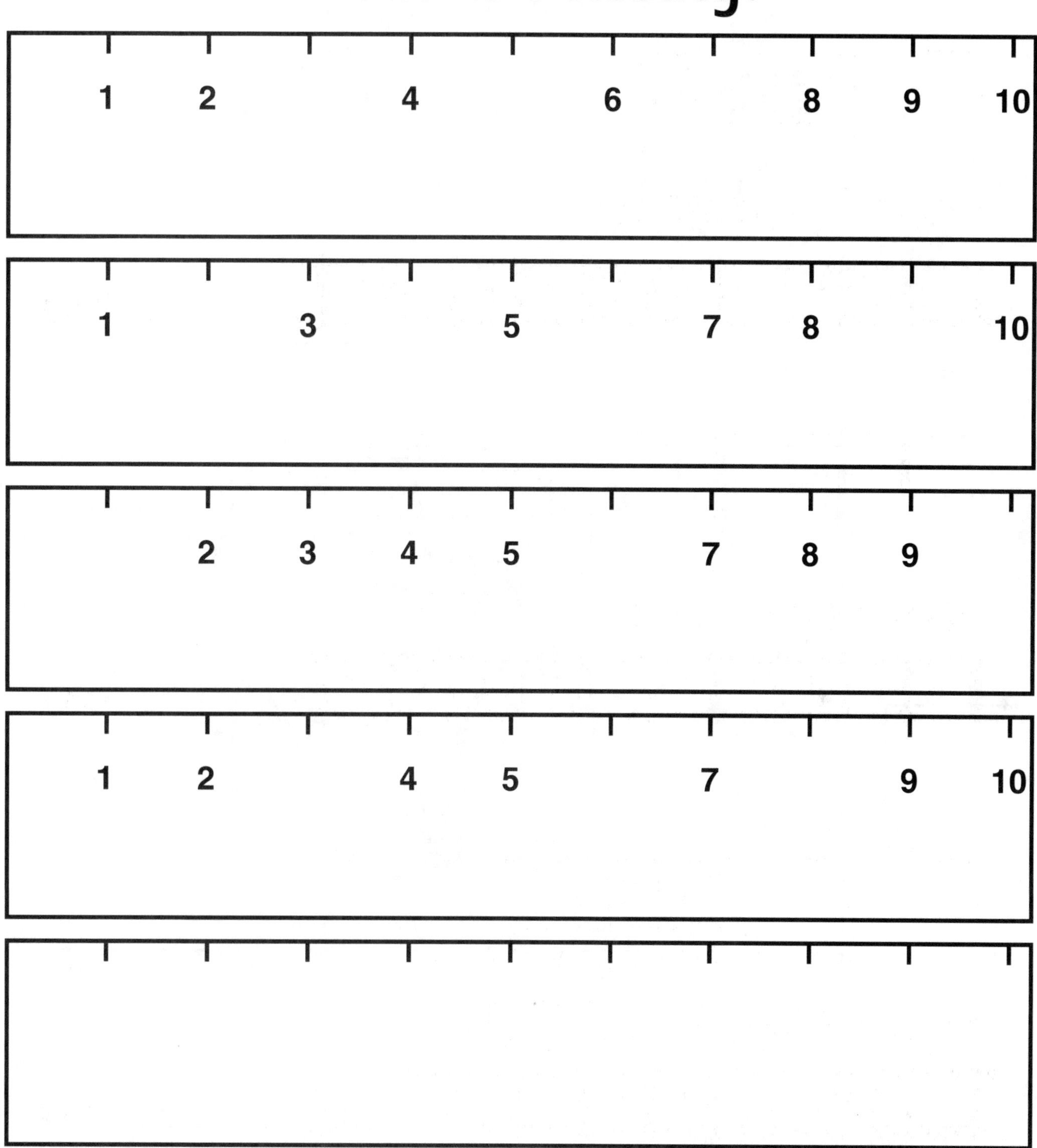

Directions: Rulers tell people the length of objects. Fill in the missing numbers on the first four rulers. Fill in all the numbers on the last ruler.

Name ____________________

Ruler Fun

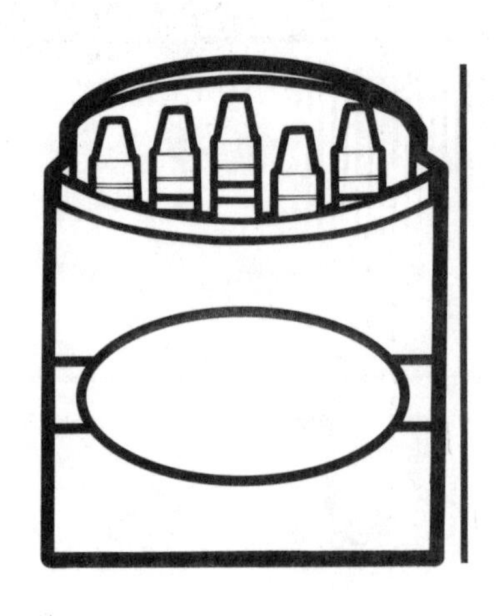

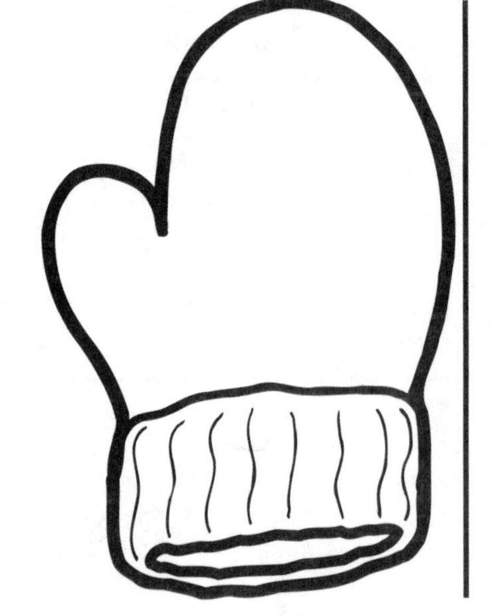

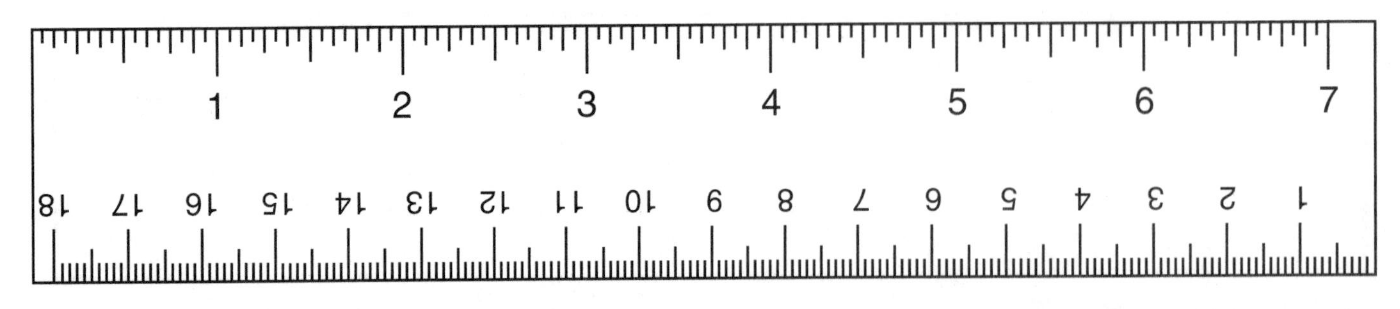

Directions: Cut out the ruler. Measure the objects above. In each box, write how long each object is.

Name ______________________________

Clown Around

✎ •

Directions: Look at the two clowns at the top of the page. Draw a line to the pants at the bottom that will fit each clown. Color the pictures.

Name ____________________

Estimate

	Estimate	Actual Amount
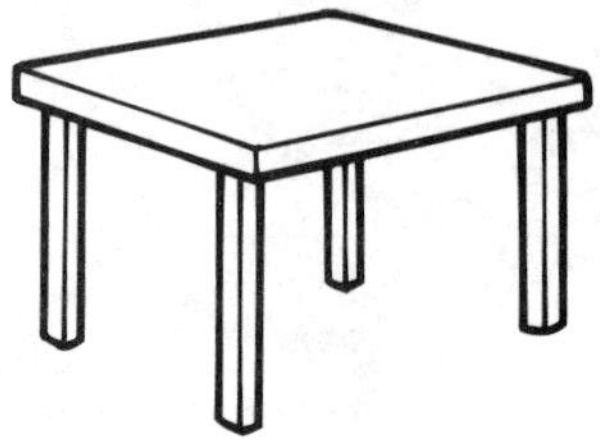		

Directions: Use your shoe as a measuring tool to measure real objects around your house or in the classroom. First, estimate how many shoes long each item will be. Write that number under the "estimate" column. Next, measure each item with your shoe. Write the length in the "actual amount" column.

Name ____________________

Say Cheese!

Directions: Use a yellow crayon to color the piece of cheese in each row that has the *most* holes. Cross out the cheese with the *fewest* holes in each row.

Name ____________________

Pizza Parlor

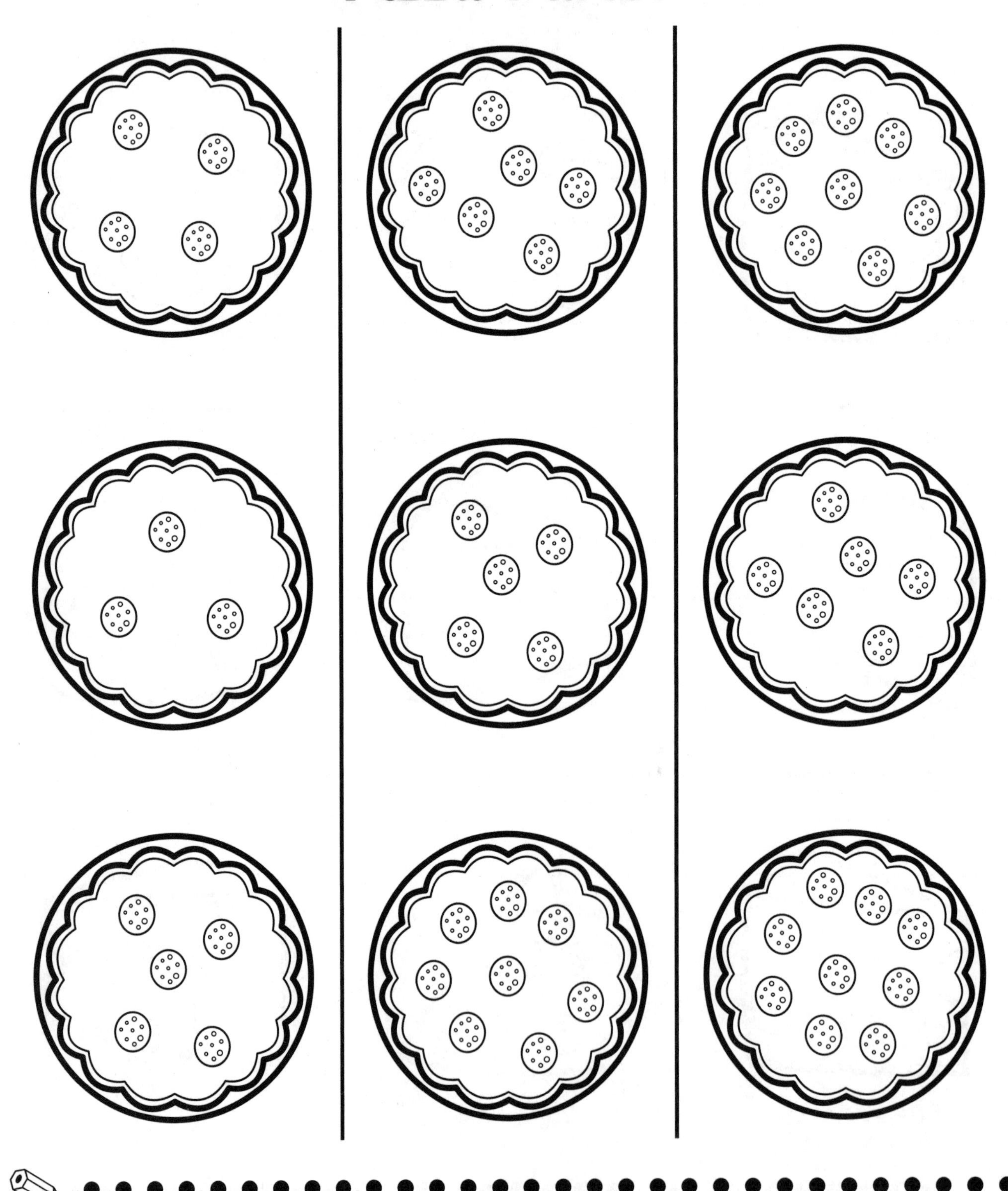

Directions: Look at the pizzas in each column. Color the pizza in each column that has the most pieces of pepperoni. Cross out the pizza with the fewest pieces of pepperoni.

Name____

Which Holds the Most?

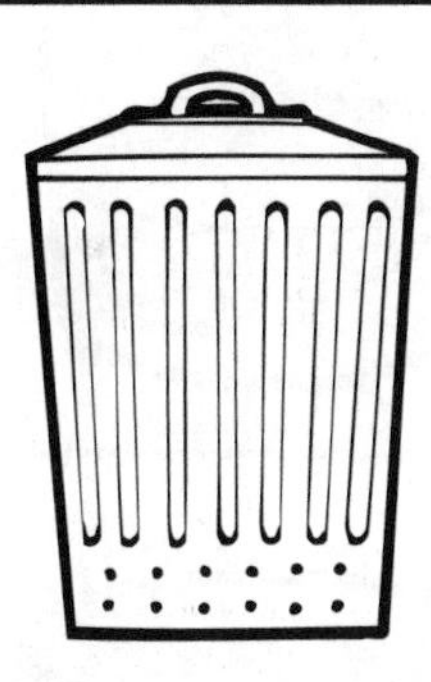

Directions: Look at each container. Color the container in each row that would hold the most.

Name ____________________

Got Milk?

Directions: Look at each row. Color the milk brown in the cup that is *full*. Cross out the cup in each row that is *empty*.

Name ____________________

How Many Fit?

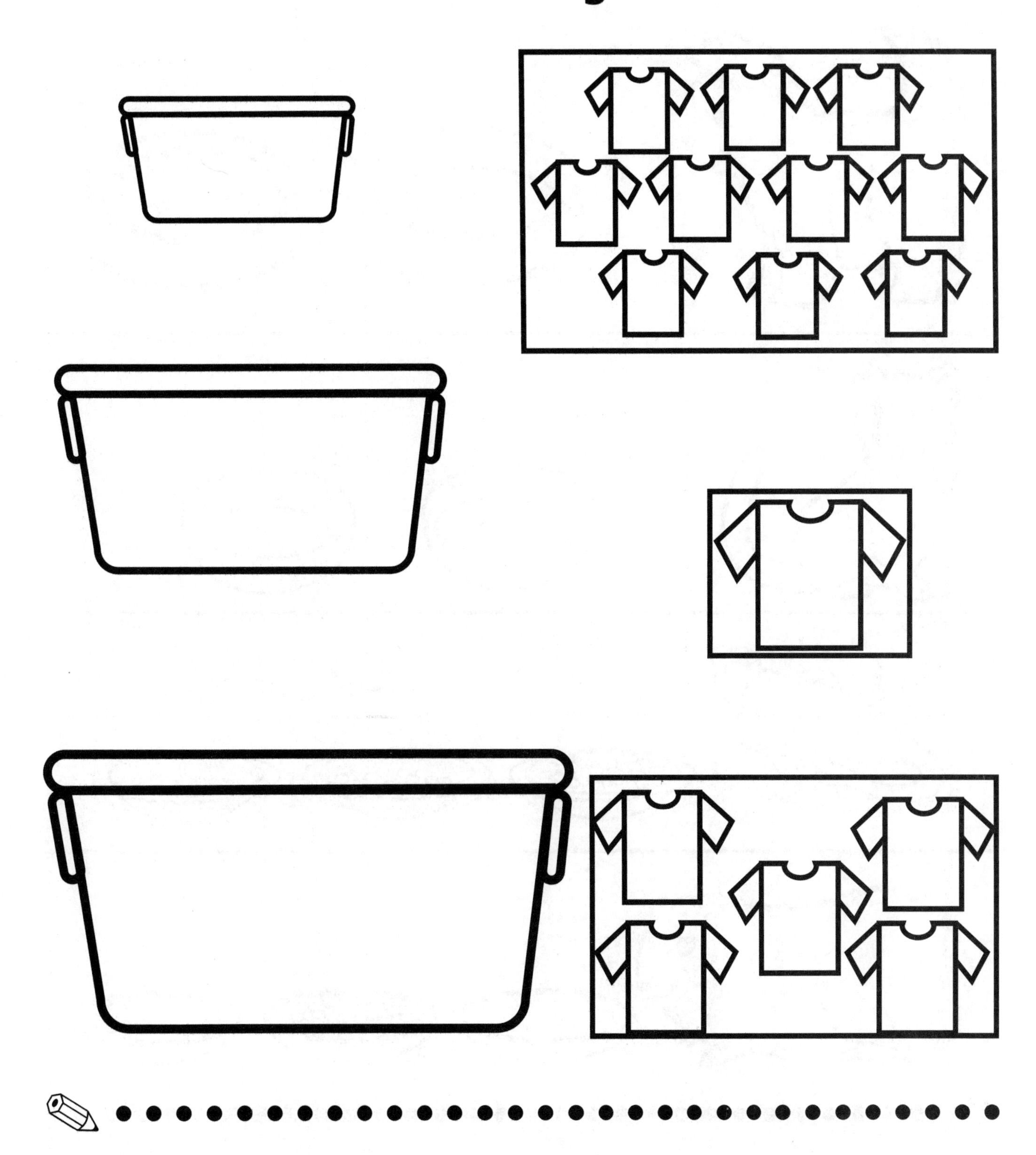

Directions: Look at the baskets. Draw a line to the set of objects that will fit in each basket. Which basket will hold the most shirts?

Name ________________________________

Who's Got More?

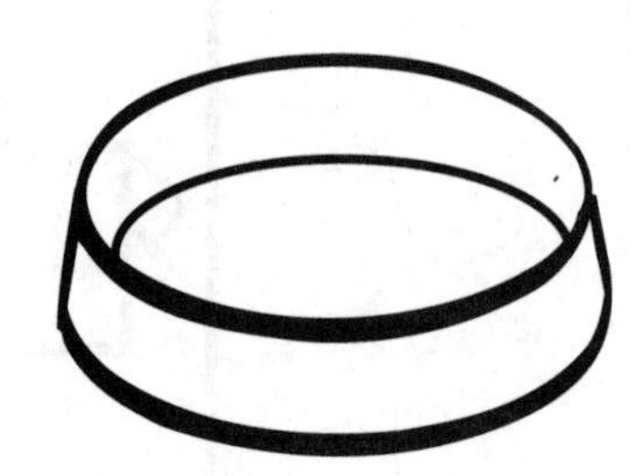

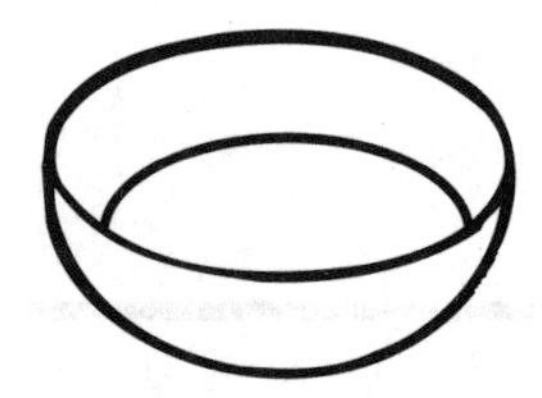

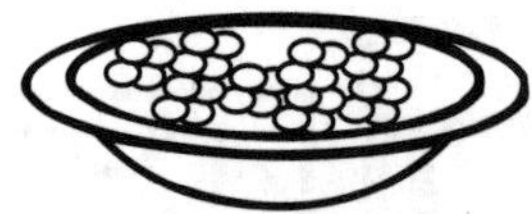

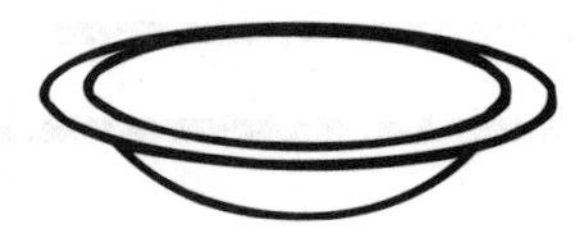

Directions: Look at each of the pet's bowls. Circle each animal's bowl that has the most food.

Name____

Three Men in a Tub

Yes No Yes No

Directions: Look at the three men and the bathtubs. Circle *yes* if all three men will fit in the tub and *no* if they will not fit in the tub.

Name ______________________________

Fly Away!

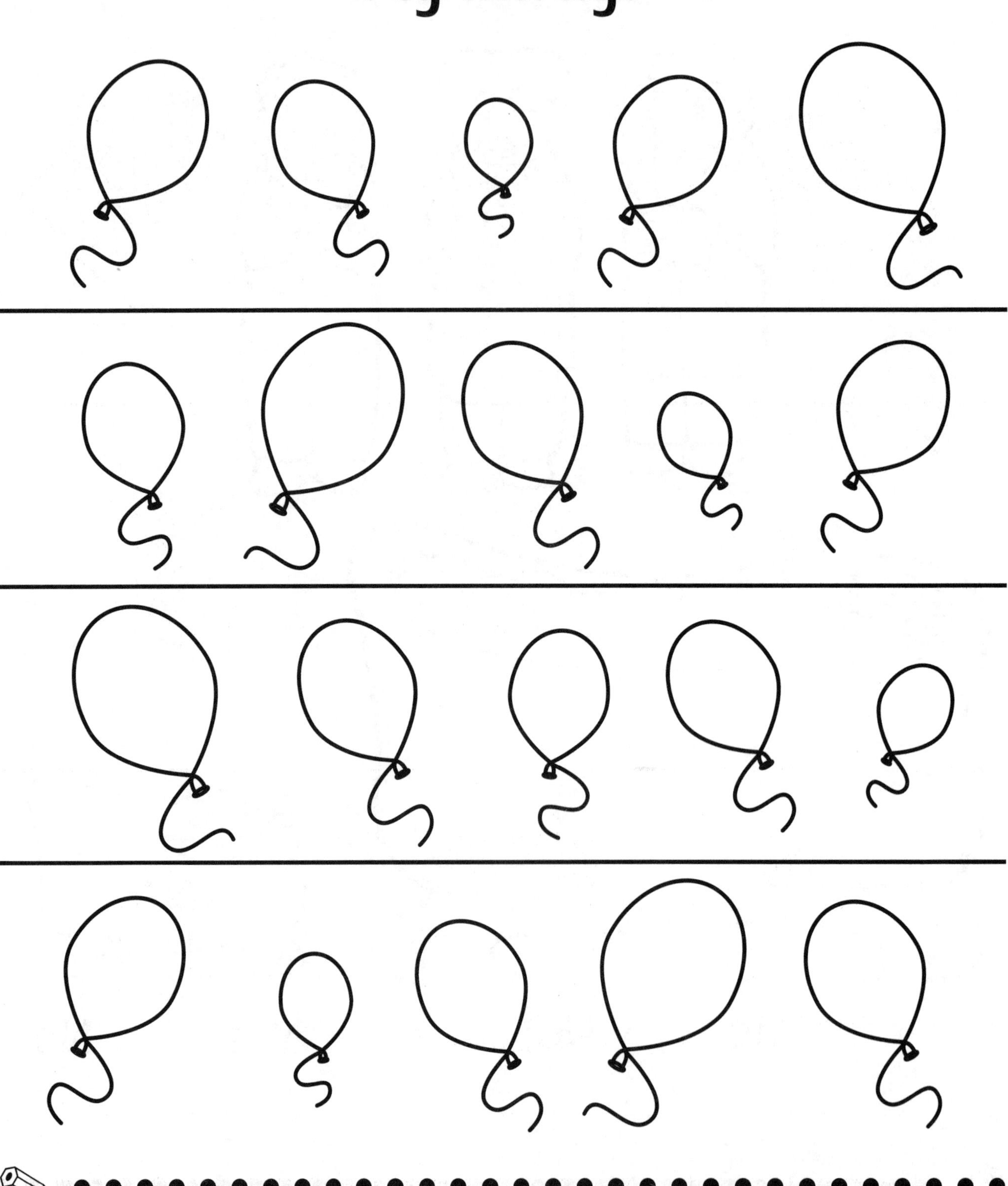

Directions: Color the largest balloon in each row red. Color the smallest balloon in each row blue. Color the other balloons green.

Name ______________________________

Small, Medium, Large

_______ _______ _______

_______ _______ _______

_______ _______ _______

Directions: Label the pictures above with the letters "S" for *small*, "M" for *medium,* and "L" for *large*.

Name ____________________

Heavy Load

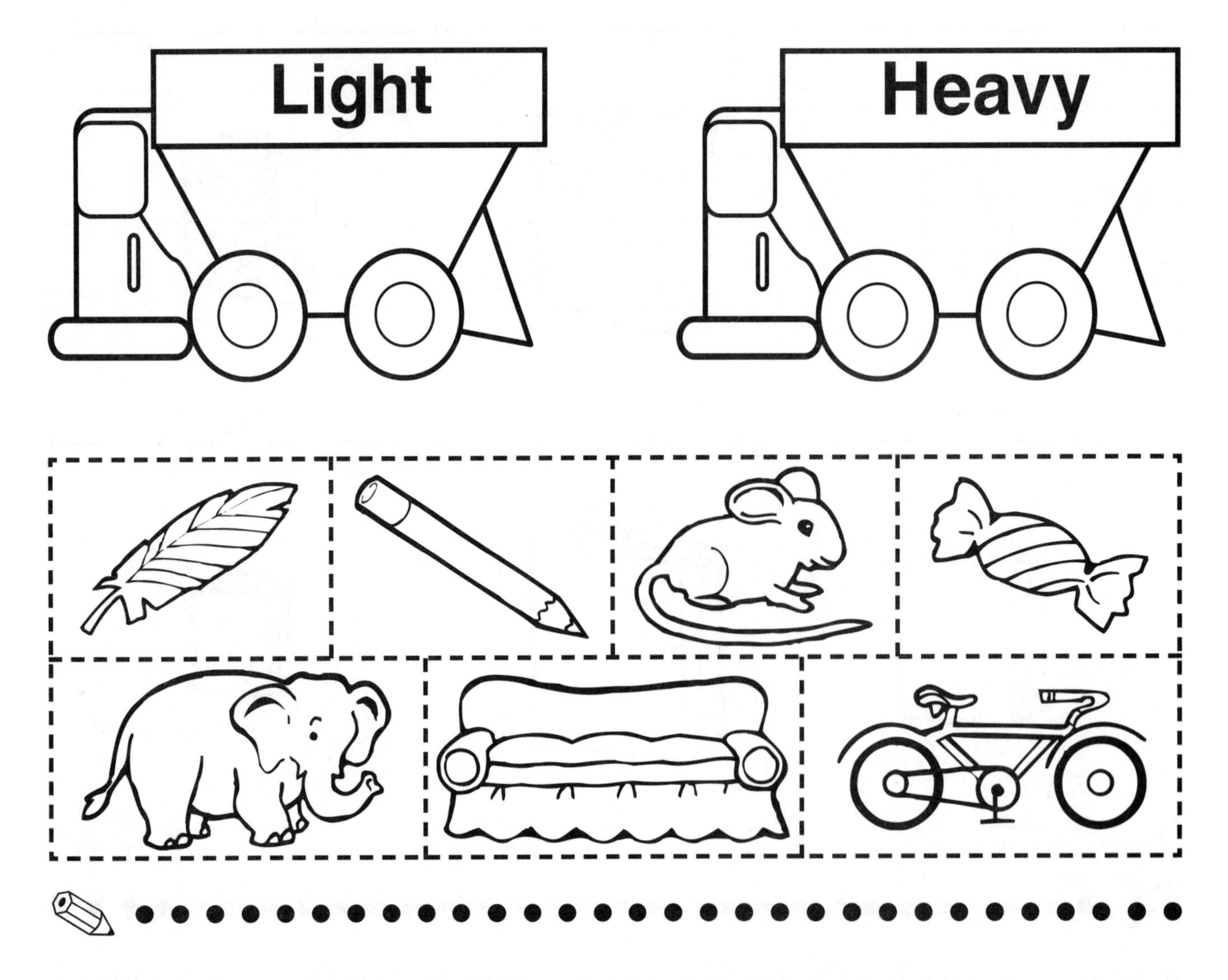

Directions: Cut out the objects in the boxes. Glue all the heavy objects in the heavy dump truck. Glue all the light objects in the light dump truck.

Name________________________________

Which is Heavier?

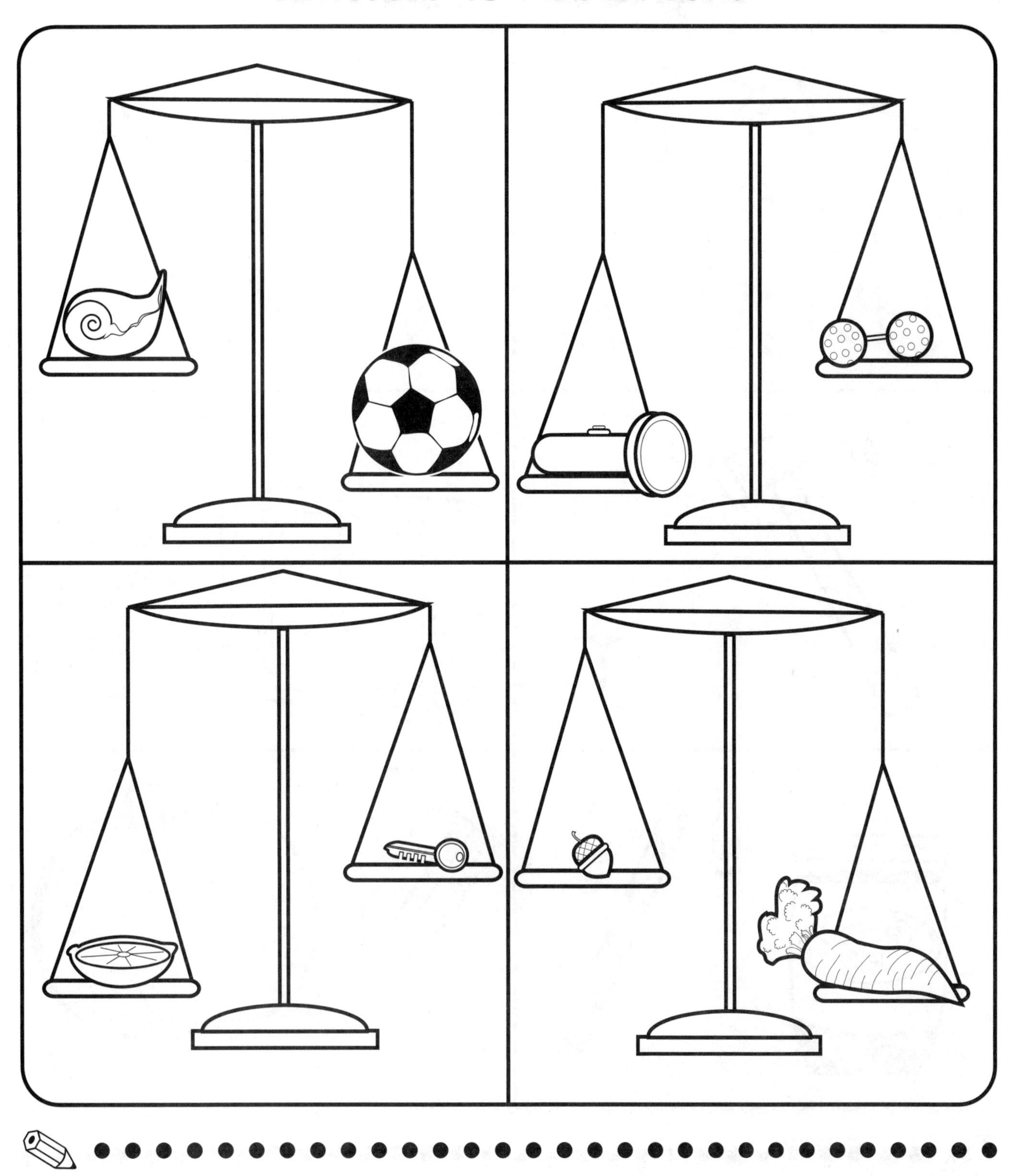

Directions: Look at the balancing scales above. Circle the item that is heavier on each scale.

Name____

Light as a Feather

Directions: Circle the item that is lighter in each box. Color the lighter item.

Name ______________________________

About the Same

 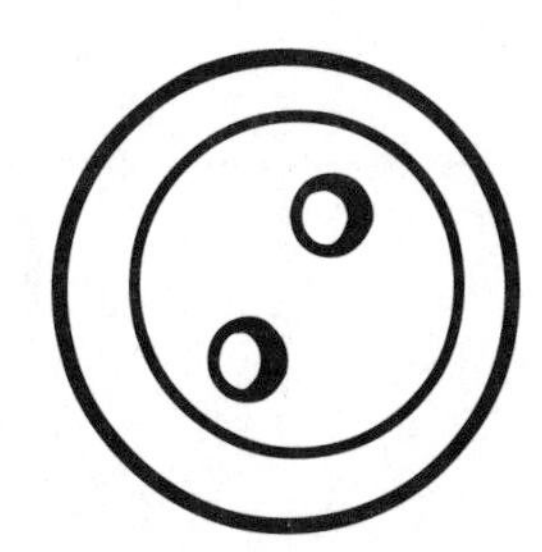 

Directions: Look at the pictures above. Think about how much they would weigh if they were real. Color the two items in each set that would weigh about the same.

Name________________________________

How Much Do You Weigh?

______ pounds

______ pounds

______ pounds

______ pounds

Directions: Look at the numbers on the scales to see how many pounds each object weighs. Write the weight on the lines below the pictures.

